DISCARD

EXPLORING THE ELEMENTS

Iron

Henrietta Toth

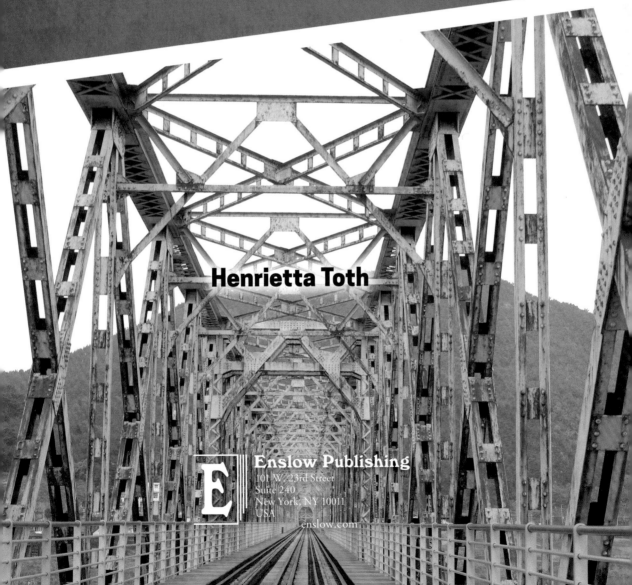

Enslow Publishing
101 W. 23rd Street
Suite 240
New York, NY 10011
USA
enslow.com

Published in 2019 by Enslow Publishing, LLC.
101 W. 23rd Street, Suite 240, New York, NY 10011

Copyright © 2019 by Enslow Publishing, LLC.

All rights reserved

No part of this book may be reproduced by any means without the written permission of the publisher.

Library of Congress Cataloging-in-Publication Data
Names: Toth, Henrietta, author.
Title: Iron / Henrietta Toth.
Description: New York, NY : Enslow Publishing, LLC, [2019] | Series: Exploring the elements | Audience: Grades 5 to 8. | Includes bibliographical references and index.
Identifiers: LCCN 2017049926| ISBN 9780766099142 (library bound) | ISBN 9780766099159 (pbk.)
Subjects: LCSH: Iron—Juvenile literature. | Transition metals—Juvenile literature. | Chemical elements—Juvenile literature.
Classification: LCC QD181.F4 T68 2018 | DDC 546/.621—dc23
LC record available at https://lccn.loc.gov/2017049926

Printed in the United States of America

To Our Readers: We have done our best to make sure all website addresses in this book were active and appropriate when we went to press. However, the author and the publisher have no control over and assume no liability for the material available on those websites or on any websites they may link to. Any comments or suggestions can be sent by email to customerservice@enslow.com.

Portions of this book appeared in *Iron* by Heather Hasan.

Photo Credits: Cover, p. 1 (chemical element symbols) Jason Winter/Shutterstock.com; cover, p. 1 (bridge) rainbow4611/Shutterstock.com; p. 5 Dorling Kindersley/Getty Images; p. 9 Bjoern Wylezich/Shutterstock.com; p. 10 Carlos Clarivan/Science Source; p. 13 Peter Hermes Furian/Shutterstock.com; p. 18 Levent Konuk/Shutterstock.com; p. 21 Photo Researchers/Science Source; p. 22 Alfred Pasieka/Photolibrary/Getty Images; p. 25 Levent Konuk/Shutterstock.com; p. 26 vilax/Shutterstock.com; p. 29 Alfred Pasieka/Science Source; p. 32 MemoPlus/Shutterstock.com; p. 35 GIPhotoStock/Science Source; p. 38 bitt24/Shutterstock.com; p. 41 Triff/Shutterstock.com.

Contents

Introduction 4
Chapter 1: Iron: An Abundant Metal 8
Chapter 2: Iron's Physical and Chemical Properties 17
Chapter 3: Finding Iron 24
Chapter 4: Compounds Containing Iron 31
Chapter 5: Iron in Daily Life 37

Glossary 43
Further Reading 45
Bibliography 46
Index 48

Introduction

Skyscrapers of varying heights top the skylines of most modern cities. This would not be possible without the element and metal called iron. Iron was first used in building a skyscraper in New York City in 1848. Today, steel makes it possible for engineers to construct even taller buildings, such as One World Trade Center in lower Manhattan. Iron is used in the production of steel, which is an alloy of iron and carbon. Steel is stronger yet lighter in weight than iron. The massive tower cranes that piece together skyscrapers are also manufactured from steel.

Iron is one of the most abundant metals on the earth. It has been used for thousands of years.

Archaeologists have found tools and jewelry that the ancient Egyptians fashioned from iron mixed with smaller amounts of other metals. Iron can also be found deep within the earth. The planet's inner core consists of an alloy of iron and nickel.

INTRODUCTION 5

Iron is present every day in some way, shape, or form.
It's in the meteorite shooting across the sky, and
it's in the bills spit out by an ATM machine.

Iron is present every day in some way, shape, or form. It's in the meteorite shooting across the sky, and it's in the bills spit out by an ATM machine. Yes, even money contains iron! It's in the ink used on the bills.

Iron plays a large part in daily life, from building skyscrapers to building red blood cells. The human body's metabolism benefits from eating the right amount of iron. Iron occurs naturally in some foods, such as meat, dairy products, and green leafy vegetables. It is added to other foods, especially breakfast cereals.

Iron is also a chemical element. Its symbol is Fe on the periodic table, a format used for organizing all the elements by atomic number and mass. By itself iron is a rather soft metal. When other elements are added to iron, they increase its hardness. This is how steel is produced.

An important and fascinating aspect of iron is its magnetic property. There are hundreds of uses for magnets in many different fields and industries. Magnets are a component in computers and other electronics like television screens and speakers. Electric generators turn mechanical energy into electricity with the help of magnets. Medical equipment such as magnetic resonance imaging (MRI) machines take detailed pictures of the human body's bones and organs. Magnets are at work in homes, too. Magnetic principles drive blenders, vacuum cleaners, and washing machines. Magnets also keep refrigerator doors closed!

One of the most important uses of a magnet is in a compass. The earth acts like a magnet because its core is made mostly of iron. The magnetism is strongest at the earth's north and south poles. A compass will always point in the direction of the North Pole. The magnet in the compass lines up with the magnetic field at the top of the earth.

The element and metal iron is billions of years old. Despite its long history, iron is still one of the most used and relied upon metals today.

1

Iron: An Abundant Metal

The word "iron" comes from the Anglo-Saxon language. Iron's chemical symbol, Fe, comes from the Latin word for iron, *ferrum*. Iron is the metal on which modern civilization is built. In its pure form, iron is used in tools, in washing machines, and in cars. Pure iron is a shiny silver-colored metal.

Unfortunately, pure iron rusts very quickly when exposed to air and water. Iron is more useful when it is combined with other substances, such as nickel (Ni), carbon (C), chromium (Cr), and titanium (Ti), to make steel. Steel is used to make many things that are used every day, from paper clips to skyscrapers. Iron is one of the most important and abundant elements in the world.

A meteorite, showing off its shiny iron

Elements as Building Blocks

Everything in the universe is composed of one or more elements. An element is a substance that cannot be broken down into smaller parts. Each element is made up of only one kind of atom, and each atom of iron is exactly the same. Atoms are very tiny. It would take 200 million of them, lying side by side, to form a line only 0.4 inches (1 centimeter) long! Amazingly, atoms are made up of something even smaller—subatomic particles.

Parts of an Atom

There are three major subatomic particles that make up the atom: neutrons, protons, and electrons.

Neutrons and protons are clustered together at the center of the atom to form a dense core called the nucleus. Neutrons carry no electric charge, while protons have a positive electrical charge. This gives the nucleus an overall positive electrical charge. Iron

The atomic structure of iron

has twenty-six protons in its nucleus, so its nucleus has a charge of +26.

Arranged in layers around the nucleus of an atom are negatively charged electrons called shells. The electrons are not fixed in a single position but orbit around the nucleus. The negative electrons are attracted to the positive nucleus. It is this attraction that holds the electrons around the nucleus. The positive and negative charges of the atom balance out, meaning the number of protons and electrons are equal. Therefore, since iron has twenty-six protons, it also has twenty-six electrons.

Organizing the Elements

There are more than one hundred known elements. As more elements were discovered over the years, they had to be organized. Scientists arranged the elements on a big chart called the periodic table.

Iron at a Glance

Chemical symbol: Fe
Atomic number: 26
Atomic mass: 55.845
Protons: 26
Electrons: 26
Neutrons: 30
Phase at room temperature: Solid
Density: 7.86 g/cm³ at 293 K and 1 atm
Melting point: 1,535° C (2,795° F, 1,808 K)
Boiling point: 2,750° C (4,982° F, 3,023 K)
Classification: Transition metal
Color: Silver
Discovered by: Was known by the ancients

The periodic table used today is based on the work of Russian chemist Dmitry Mendeleev. He published the first version of the table in 1869 while teaching chemistry at the University of St. Petersburg in Russia. Mendeleev sought to organize the elements to make it easier for his students to study and understand them. He arranged the elements in horizontal rows, according to weight, with the lightest element of each row on the left end and the heaviest on the right. Though Mendeleev's periodic table did not list all of the elements that are known today, iron was among those that he included.

Today, the elements on the periodic table are listed in order of increasing atomic number, which is the number of protons. Arranged like this, there are many trends, or patterns, that can help to classify the elements. The location of an element on the periodic table predicts whether it is a metal, a nonmetal, or a metalloid. Metals can be recognized by their physical traits. Generally, metals can be polished to be made shiny. Also, metals conduct electricity. Most metals can also be hammered into shapes without breaking. This is called malleability. Metals are usually ductile, which means that they can be made into wires. Substances such as plastics, glass, and wood are classified as nonmetals because they lack the characteristics of metals. Metalloids, or semimetals, have characteristics of both metals and nonmetals. In most respects, they behave like nonmetals. However, they are able to conduct electricity, though not nearly as well as metals do.

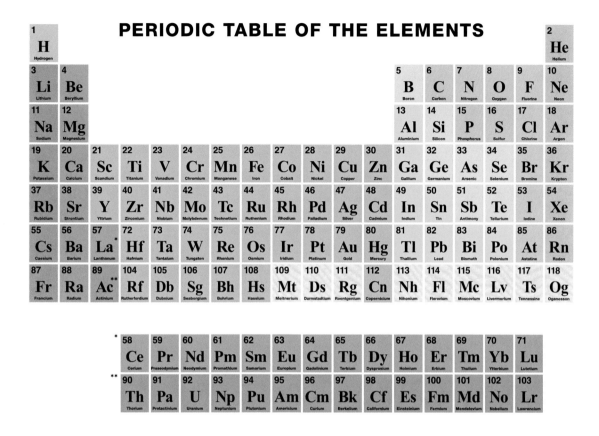

The periodic table of elements is organized into groups and periods.

The elements on the periodic table are divided by a "staircase" line. The metals are found to the left of this line and the nonmetals are on the right. Most of the elements bordering the staircase line are metalloids. Iron, as a metal, is located to the left of the staircase line.

The Periodic Table Setup

The periodic table is very useful because it shows a lot about an element just by its location on the table. The horizontal line of elements from left to right is called a period. Elements are arranged in periods based on the number of electron shells that surround the nucleus of their atoms. The outermost electrons determine how elements behave and react with one another. Iron is in period IV.

Going down the chart from top to bottom, the vertical lines of elements are called groups or families. The elements in a given group have similar characteristics called properties. Iron is in group VIII, which consists of nine elements and is located in the middle of the periodic table. The elements found in group VIII have the typical properties of metals, such as malleability and the ability to conduct electricity. Within this group, horizontal similarities are greater than vertical ones. Therefore, each horizontal set of elements is thought of as a triad. Iron, cobalt (Co), and nickel have similar properties, so they have been grouped together to form the iron triad.

Why Elements Are Different

What makes iron different from other elements like oxygen (O) and silver (Ag)? The difference is the number of protons found in the nucleus of the atoms. The number of protons makes each element unique, so they are organized by these numbers on the periodic table. The number of protons that are found in an atom of an element is called the atomic number. On the periodic table, this number is found above the element's symbol. Iron has twenty-six protons, so its atomic number is 26. The fact that iron has twenty-six protons in its nucleus is what makes it iron. If one proton were added to iron's nucleus—giving it twenty-seven protons—it would be an entirely different element. Adding another proton would create the atom of the element cobalt. Taking away one of iron's protons results in manganese (Mn), which has twenty-five protons in its nucleus. Changing the number of protons creates an entirely different element.

Calculating Neutrons

The number that is found below an element's symbol on the periodic table is called the atomic mass or atomic weight. Iron has an atomic mass of 55.845. Knowing the atomic mass of an element helps to figure out how many neutrons there are in an atom of that element. The atomic mass is the sum of the number of protons and neutrons in the atom. Therefore, knowing that the atomic mass of iron is 55.845 and the atomic number (number of protons) is 26, it is possible to figure out how many neutrons there are in an atom of iron by subtracting the two numbers:

$$55.845 - 26 = 29.845$$

Iron has approximately thirty neutrons in its nucleus. A lot of information about iron can be found out just by looking at the periodic table!

Multiple Valences

Iron and the other group VIII elements are more generally classified as transition metals. They are located in the middle of the periodic table, from group IIIB on the left to group IIB on the right. Transition metals are unique because many of them are able to do something that other elements cannot: they are able to form more than one positive ion. An ion is a charged particle. It is formed when an atom gains or looses electrons from the shells that surround the nucleus. Atoms are usually electrically neutral, which means they carry no charge. This is because they have an equal number of positively charged protons and negatively charged electrons. Basically, the charges cancel out each other. However, if an atom picks up extra electrons, it becomes a negatively charged ion (called an anion). In the same manner, if an atom loses electrons, it becomes a positively charged ion (called a cation). Most elements form only one ion by loosing or gaining a certain number of electrons. This is not true of the transition metals. Iron, for example, has the ability to form two different ions, Fe^{2+} and Fe^{3+}. Fe^{2+} results when an iron atom loses two electrons, and Fe^{3+} is formed when an iron atom loses three electrons.

2

Iron's Physical and Chemical Properties

All elements have characteristic physical and chemical properties. These properties help scientists identify and classify the element. The physical properties of an element are those that can be observed without changing the element's identity or properties. Some examples of physical properties are an element's phase at room temperature (whether it is a solid, liquid, or gas), hardness, melting point, and color. The chemical properties of an element describe its ability to undergo a chemical change. A chemical change converts one kind of matter into a new kind of matter. An example of a chemical property is iron's ability to react with oxygen to form rust.

Natural iron ores

The chemical properties of an element cannot be observed without changing the identity of the element. For example, in the process of observing how iron rusts, iron is changed from a shiny metal to a crumbly, reddish-brown material.

Physical States

At room temperature, an element is found in one of three physical states: solid, liquid, or gas. Knowing the physical state, or phase, of an element at room temperature helps scientists to identify it. Iron is found in the solid phase at room temperature. A solid has a fixed shape and volume. Solids resist being compressed or having their shape changed.

Mass per Volume

The density of iron is 7.86 grams/centimeter^3. Density measures how compact an object is—that is, how much mass it contains per unit volume. Solids are denser than liquids, which are, in turn, denser than gases. If iron shavings are sprinkled on water, they will sink because iron is more dense than water.

How Hard Is Iron?

When iron is in its pure form—meaning it is not mixed with anything else—it is a relatively soft metal. On Moh's hardness scale, iron has a hardness of 4. For comparison, a fingernail has a hardness of 2.5 and a penny has a hardness of 3.5. Iron is harder than both of

Determining Iron's Hardness

Moh's hardness scale is used to classify the hardness of minerals, metals, and other products. This scale was published in 1822 by Frederick Mohs, an Austrian mineralogist. He got the idea for the scale from miners, who routinely performed scratch tests. The scale shows ten groups of substances, in order of increasing hardness. Each successive substance is able to scratch the preceding ones and can be scratched by all that follow it. For example, iron is able to scratch copper, but it can be scratched by hardened steel.

Hardness Rating	Examples
1	Talc
2	Gypsum, rock salt, fingernail
3	Calcite, copper (Cu)
4	Fluorite, iron
5	Apatite, cobalt
6	Orthoclase, rhodium (Rh), silicon (Si), tungsten (W)
7	Quartz
8	Topaz, chromium (Cr), hardened steel
9	Corundum, sapphire
10	Diamond

these and could scratch them. However, there are lots of materials that are able to scratch pure iron. In order to make iron harder, other elements are added to it. This is how steel is made, which is much harder than pure iron!

Conducting Electricity and Heat

Like all metals, iron conducts electricity and heat. This means that heat and electrical current are able to move through iron. Metals are able to conduct electricity because the electrons from the outer shells of its atoms are able to move about in a "sea" of electrons. As these electrons move, they carry the electricity with them. The sea of electrons that metals have also makes them good conductors of heat. When metals, such as iron, are heated, the electrons gain more energy. This makes them move about more quickly and distribute the heat throughout the whole metal.

It takes a lot of heat to pull apart iron atoms. Iron's melting point is 1,535° C (2,795° F, 1,811 K). Because solid iron can handle a lot of heat without melting, it is often used to build things that undergo intense heat, such as a car engine.

Once iron is melted into a liquid, the temperature must reach a steamy 2,750° C (4,982 °F, 3,023 K) before it will boil and become a gas. To see how hot that is, compare this temperature to the temperature found on the surface of the sun: 5,500° C (9,932° F, 5,773 K). Iron would definitely boil there!

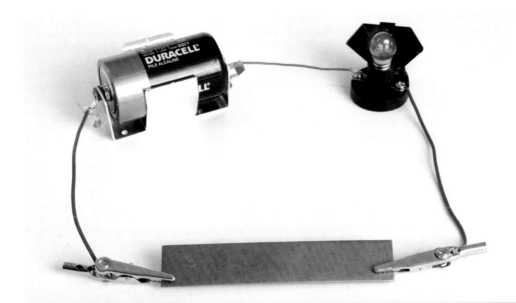

Here, iron conducts electricity from a battery to light a bulb.

Iron's Magnetism

One of iron's most important properties is its strong attraction to magnets. Other metals, such as nickel and cobalt, are also magnetic, but iron is much more common than those elements. Finding out if an object contains iron is easy by testing it with a magnet. Magnetism is a force that either pulls things together or forces them apart.

The word "magnetism" comes from the ancient Greek city of Magnesia, a place where lodestones were found in ancient times. Lodestones are rocks that contain iron and were the first known magnets. Although these rocks had become permanently magnetized, most iron is not permanently magnetic. Iron can become

Iron particles stand straight up on a magnet.

magnetic because of the structure of the iron atom. The electrons orbiting the nucleus of the iron atom make each atom act like a tiny magnet. If the electrons are lined up in just the right way, a piece of iron becomes magnetized. Iron is made magnetic when it is properly exposed to a magnetic field. A magnetic field can be produced by another magnet or by an electric current. Heating or melting the iron, however, can make the metal lose its magnetism.

The earth is like a huge magnet because its core is made mostly of iron. The core is so hot that the outer part is always liquid. Scientists think that, as the earth spins, the liquid iron swirls around. This movement creates the earth's magnetism.

The earth has magnetic poles, just like a bar magnet. The earth's magnetism is the strongest at these poles. If a small magnet is allowed to turn freely, it will point toward the earth's poles. That is how a compass works—the magnet inside points toward the direction of the North Pole.

Iron's Appearance in Nature

Iron, in its pure form, is a somewhat shiny, silvery white metal. However, iron is rarely found in its pure state in nature. This is because, chemically, iron is a fairly reactive metal, meaning it easily combines with other elements. Iron atoms easily lose two or three of their electrons when they are exposed to other elements like oxygen. When an iron atom loses two or three electrons to some other element, it stops being shiny and takes on a black, green, yellowish orange, or bright red color. This is a chemical change because the identity of iron has changed, such as when rust occurs. Here, the iron atoms have bonded with oxygen atoms and water molecules to turn iron into rust.

3
Finding Iron

Iron is believed to be the most common metal in the universe. As a gas, it is found in the sun and in many other stars. The earth's core is made mostly of iron. Iron is also the fourth most common element and the second most abundant metal in the earth's crust. Although iron makes up about 5 percent of the earth's crust, hardly any of it is in its pure form. The only pure iron that is found on the earth's surface arrives from outer space in the form of meteorites. However, a lot of iron can be found combined with other elements.

Iron Is Unrefined

Iron ores are rocks that contain iron and other elements. Iron was concentrated in these rocks by natural forces during the formation of the earth's crust. People have been trying to extract iron from these rocks for thousands of years, and it is these ores that

Ores often look red because of the iron inside of them; it's the color of rust.

provide the iron used today. Unlike in ancient times, the ability to get iron from ore has made iron a very commonly used metal. Because iron is so common, it is one of the world's cheapest but most useful metals.

The main ores that provide iron are hematite, magnetite, and pyrite. These iron ores are mined. Hematite and magnetite are the richest in iron, containing about 70 percent of the metal. These ores contain iron and oxygen and are called iron oxides. Hematite may be black, brownish red, or dark red. Magnetite is black and has magnetic properties. Pyrite is about half iron and half sulfur (S). Pyrite is commonly called fool's gold. Its shiny gold-colored appearance has fooled many people into thinking that they have found gold! Iron is also extracted from the limonite, siderite, and taconite ores.

Different Types of Iron

There are many different types of iron, but all of these can be placed into three groups: cast iron, wrought iron, and pig iron. Cast iron is an iron alloy that contains iron and two other elements: carbon and silicon. Cast iron is very hard because it contains carbon. However, cast iron is also very brittle, which means it breaks apart easily. It cannot be shaped with a hammer even if it is heated to a very high temperature; it will simply break apart. Cast iron is made into useful things when it is melted and then allowed to cool in a mold. The liquid iron takes the shape of the mold, and when it cools and hardens it keeps that shape. Cast iron is used in manhole covers, automobile parts, and cookware.

Pig iron is used for manhole covers in the city.

Wrought iron is nearly pure iron. It is mixed with just a tiny bit of a glasslike material, called iron silicate. Unlike cast iron, wrought iron is malleable. This means that it can be hammered into different shapes. Wrought iron also resists corrosion, or rusting, better than does cast iron. Because it is malleable, wrought iron is used to make metal fences with fancy designs, coat racks, and handrails on staircases.

Pig iron is made in a blast furnace. It contains some carbon and a small amount of other elements. At one time, people took the liquid pig iron from the blast furnace and poured it into molds. Pig iron got its name because the molds that it was poured into looked like a group of baby pigs gathered around their mother. Today, most pig iron is used to make steel.

Steel Production

Although some elemental iron is made into iron products, most of it is used to make steel. Think of steel as refined or purified iron. Unlike pure iron, steel is not an element. Steel is produced by refining iron and then mixing it with other elements. Making steel involves first removing unwanted substances, such as excess carbon, silicon, or sulfur. To do this, the iron is mixed with limestone and heated to a very high temperature. This is normally done in one of three furnaces: the open hearth, the electric, or the basic oxygen. Once this is done, the desired materials are carefully added, which make steel a lot stronger than iron.

Four Groups of Steel

There are thousands of kinds of steel that can be grouped into four types: carbon steel, alloy steel, stainless steel, and tool steel. Carbon steel is used more often than any other kind of steel. The properties of this steel depend on how much carbon it contains. The higher the amount of carbon, the harder the steel. Carbon steel is made into many products, such as structural beams in buildings, automobile bodies, kitchen appliances, and cans.

Alloy steel contains some carbon but its properties depend on the other elements it contains. Each element that is added to alloy steel improves one or more of its properties. Nickel, for example, makes steel tougher. When manganese is added, it makes steel harder, tougher, and more resistant to wear. Molybdenum (Mo), another element, increases steel's hardness and resistance to corrosion, or rusting. Other elements used in alloy steel include chromium, aluminum (Al), tungsten, copper, titanium, silicon, and vanadium (V).

Stainless steel contains a large amount of the element chromium. Many stainless steels also contain nickel. This type of steel resists corrosion better than any other steel. It is used in automobile parts, hospital equipment, razor blades, and pots and pans.

Tool steel is a very hard steel. It is made by a process called tempering. In this process, certain types of carbon steel and alloy steel are heated to a high temperature and then quickly cooled. Tool steel is used in metalworking tools such as files, drills, and

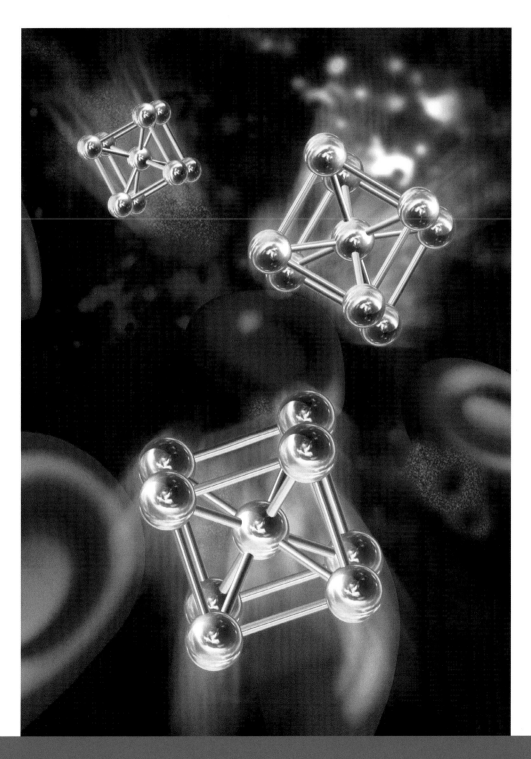

The red blood cells in your body contain iron!

chisels. Tools made from tool steel are then used to cut other softer metals.

Red Blood Cells and Iron

Iron is important not only in bridges, skyscrapers, and automobiles, but also in the human body. Iron is a part of hemoglobin, which is in red blood cells. Hemoglobin is responsible for carrying oxygen from all the parts of the body. Each hemoglobin molecule contains four iron atoms. Blood and rust are red for the same reason—because of iron's strong attraction to oxygen. When people breathe, oxygen from the air enters their lungs. Here, it is attracted to the iron in hemoglobin and combines with it to form oxyhemoglobin, the specific hemoglobin that contains oxygen. The oxyhemoglobin is then transported throughout the body red blood cells, and oxygen is released wherever it is needed.

A lack of iron in the body is known as iron deficiency. Someone who does not have enough iron will be tired and will not have much energy. Iron deficiency can be due to a diet that is lacking in iron or to an insufficient number of red blood cells. If a person does not have enough red blood cells, he or she is said to be anemic.

Calculating Iron Amounts in the Human Body

This formula calculates how much iron is in a person's body. Multiply the person's weight by the numbers given.

Weight in pounds x 0.02724 = the number of grams of iron in your body

Weight in kilograms x 0.06 = the number of grams of iron in your body

4

Compounds Containing Iron

An important property of metals is their ability to combine chemically with nonmetals to form what are called ionic compounds. Forming an ionic compound between a metallic element and a nonmetallic element involves the transfer of electrons from the metal to the nonmetal. It is known that iron is able to lose two or three of its electrons. This is important because it means that iron can form more than one compound with a given element. When Fe^{2+} (iron that has lost two electrons) combines with another element, the compound it forms is called a ferrous compound. When Fe^{3+} (iron that has lost three electrons) combines with another element, the compound it forms is called a ferric compound. This can be seen by the way that iron reacts with chlorine, forming two differ-

Iron + oxygen + water = rust! Rust is ferric oxide; iron loses three electrons to form it.

ent compounds. When iron loses two electrons to chlorine, it forms ferrous chloride ($FeCl_2$). When iron loses three electrons to chlorine, it forms ferric chloride ($FeCl_3$). The most well-known, and least desirable, chemical reaction of iron is its reaction with water and oxygen to form a ferric compound called rust.

What Is Rust?

Rusting requires both oxygen and water. Iron will not rust in pure water that is free of oxygen, and it will not rust in pure oxygen that is moisture free. When iron, oxygen, and water meet, rust is readily formed. Rust is the common name for ferric oxide and has the chemical formula Fe_2O_3. This is a ferric compound because iron looses three electrons to form it. It is because iron combines so readily with oxygen that pure iron is rarely seen in nature.

People do many things to keep their iron and steel possessions from rusting. One way to protect these metals is to paint them. The paint on cars forms a barrier between the iron that the car is made of and the water and oxygen in the air. Many household items have plastic coatings to protect them. A lot of storage shelves are actually wrought iron covered with plastic. Another way to protect iron and steel is to plate, or cover, it with a metal that does not rust, such as

The Red Planet Mars

Earth is not the only planet that contains iron. Other planets do, too. Mars is red because of the iron in its soil. It takes water and oxygen for iron to rust. Scientists believe that long ago Mars had these substances in its atmosphere. The iron on the surface of the planet reacted with water and formed rust. Rust is everywhere on the planet, which is why Mars is called the Red Planet.

Mars is red because of the iron in its rocks and soil.

chromium. Car bumpers are sometimes made of chrome-plated steel.

Because rust is the product of a chemical change, it cannot be separated back into iron and oxygen by any physical means, such as boiling or freezing. Recovering the iron and the oxygen from rust would require another chemical reaction. One way to get rid of rust is by "pickling" the iron. In this process, the rust reacts with an acid. This turns the rust into a harmless substance that can be washed away from the metal. Sometimes hydrochloric acid (HCl) is used. This acid reacts with the rust to form a useful iron compound called ferric chloride ($FeCl_3$).

Using Iron to Treat Water

Ferric chloride ($FeCl3$) is a dark-colored crystal. This compound is used mainly for sewage and water treatment. It helps to remove harmful impurities from drinking water by causing them to lump together. Once the unwanted material has formed a mass, it can be filtered out of the water.

Ferric chloride is also used by the food, paper, photography, and pharmaceutical industries, which need to use very pure water. Ferric chloride is also helpful for the environment. Before waste is released into rivers or lakes, it is treated with ferric chloride so that it will not harm the organisms in the water. Sometimes ferric chloride is also used as a disinfectant to destroy disease-causing bacteria and viruses.

Iron forms into chloride salts like potassium chloride. Transition metal salts are often colored.

Iron as a Mordant

Ferrous chloride ($FeCl_2$) is formed when hot chlorine gas is passed over iron. This is a ferrous compound because iron has donated two electrons to the chlorine. Ferrous chloride is a colorless, crystalline substance. Like ferric chloride, it is used to treat sewage. It is also used to dye fabric.

Ferrous chloride is known as a mordant. Mordants keep the color of fabric from fading. When dyeing something with a natural dye, which is obtained from plants or animals, it is necessary to use a mordant. If a mordant is not used, the color will fade when the fabric is exposed to sunlight or is washed. Ferrous chloride binds these natural dyes to the fabric so that they cannot be washed out.

Medical Uses of Iron

Ferrous sulfate ($FeSO_4$) is a green crystalline substance. Like ferrous chloride, it is used as a mordant. It is also used in water purification. More important, this iron compound is used to treat iron-deficiency anemia. People at a higher risk of developing this anemia include pregnant women, infants, children, and adolescents. Children need extra iron because they are growing. People whose diets are low in iron or whose bodies have trouble absorbing iron can develop anemia.

Without enough iron, red blood cells do not mature properly. They stay small and pale and are unable to carry oxygen properly. That is why an anemic person does not have the energy to complete simple tasks. Therefore, a doctor may prescribe ferrous sulfate in pill form to provide the iron that the body needs in order to make red blood cells.

5
Iron in Daily Life

Iron is an important part of the human body. That is why it is necessary to provide iron for the body to use by eating foods that are high in iron. The best source of iron is red meat. It not only contains a lot of iron, but the iron is easily absorbed, or taken in, by the body. Some other foods, such as beans and lentils, also contain a lot of iron. However, the iron from these foods is a lot harder for the body to absorb. Some foods, like fruits and vegetables, contain vitamin C, which helps the body to absorb the iron from the other foods eaten.

Dietary Iron

Iron aids the human body in many functions like breathing and metabolism. It is especially necessary for the development and growth of babies and children. Dietary iron comes from two sources: animal proteins and plant-based foods. Iron is easily absorbed by the body

Many of the foods you consume contain iron: meat, spinach, and lentils, just to name a few.

from the animal proteins in fish, meat, poultry, and seafood. Other animal products, such as eggs and dairy products like cheese and milk, also provide a fair amount of iron. Plant-based foods—like beans, fruits, grains, nuts, and vegetables—provide smaller amounts of iron that are harder to absorb.

The Institute of Medicine (IOM) sets forth the daily Recommended Dietary Allowance (RDA) of iron that is necessary to maintain good health. The RDA varies by age group and is higher for people who do not eat meat or other animal products. Vegetarians need 1.8 times more iron than the RDA. Women need more iron from ages fourteen to forty. For infants the RDA is between .27 and 11 milligrams of iron per day. Toddlers require 7 mg. Children from ages four to eight need 10 mg, while children nine to thirteen require 8 mg of iron daily. Teenagers between the ages of fourteen and eighteen should aim for 11 to 15 mg RDA. Adults from nineteen to fifty need between 9 and 18 mg daily iron. Adults over the age of fifty-one should have 8 mg RDA.

Finding Iron in Breakfast Cereal

Some foods do not naturally contain a lot of iron but have iron added to them. These are iron-fortified foods, such as breakfast cereals. Metallic iron, in the form of iron filings, is often added to cereals. These tiny pieces of iron can be separated from the cereal. First, crush the flakes of iron-fortified cereal into tiny pieces the size of a pinhead. Mix this cereal in a plastic bag with enough warm water to make it soupy. Seal the bag and press a magnet to the outside of the bag. The iron filings will actually collect on the inside of the bag near the magnet! The iron will look like small dark dots.

Monetary Magnetism

In the United States, the metal that has been used to make pennies has changed over time. In 1943, US pennies were made of steel, so these pennies were attracted to a magnet. They are now collector's items! Currently, pennies are made of copper and zinc.

Creating a Compass

It's easy to make a compass that is very similar to the ones people made hundreds of years ago!

Here are the things needed to make a compass:

- A large needle
- Something that floats (such as a cork or a plastic cap from a milk jug)
- A bowl of water
- A magnet

First, magnetize the needle. Stroke the needle with the magnet. The direction should be from the back of the needle to the point. This will make the needle magnetic. Place the float in the middle of the bowl of water and lay the magnetic needle on top. The needle and float will slowly turn so that the pointed end of the needle points north. Try turning the needle around and see what happens.

Neither of these metals is magnetic, so pennies are not attracted to a magnet.

None of the other US coins are attracted to magnets, not even nickels. Even though nickel is a magnetic metal, just like iron, nickel coins do not contain enough of the metal to make them magnetic.

Surprisingly, dollar bills are attracted to magnets! The paper is not magnetic, but the ink on the paper contains iron. The ink is magnetic. Bill-changing machines know how much change to give back because of the magnetic ink. The machines can tell the difference between the bills ($1, $5, $10, etc.) by the pattern of the ink on them. This also keeps people from making fake bills with photocopy machines. If a bill does not have the magnetic ink, the machine will recognize it as counterfeit.

How a Compass Works

A compass is a device that helps people find their way. In the center of a compass is a needle made of magnetized iron. No matter

A compass uses iron in its needle; the needle will always point to the earth's magnetic north pole.

the location, the needle on a compass always points north, toward the magnetic north pole. Although there are now more complex ways of finding direction, such as maps and GPS (Global Positioning System), people long ago relied on compasses to get around.

Surrounded by Iron

It is easy to test everyday items—like staples, coins, nails, paper clips, jewelry, and soda cans— to see if they are made of iron. The objects that are attracted to a magnet are probably made of iron.

Even though other metals are also magnetic, iron is used much more often.

Each element in the periodic table plays its role perfectly to create everything in our universe. Like all the other elements in the periodic table, iron is essential to life on the earth. Every day, every minute, every second, we are surrounded by iron. It is found in outer space and right here on earth, both on the surface and deep within the earth's core. Iron from the surface of the earth is used to make steel—one of the greatest inventions and most useful materials ever created. Iron is in the food we eat, and it is even found in us!

Glossary

acid A reactive substance that accepts electrons or donates protons. Usually has a sour, sharp, or biting taste.

alloy A mixture of a metal and one or more other elements, usually metals.

anion An atom that picks up extra electrons and becomes a negatively charged ion.

archaeologist A person who studies prehistoric cultures and people.

atom The smallest part of an element.

chemical reaction A change in which one kind of matter is turned into another kind of matter.

crust The surface layer of the earth.

crystalline Composed of crystals, with a regular pattern.

ferric compound A compound of iron and another element or elements in which the iron atom(s) has given up three electrons.

ferrous compound A compound of iron and another element or elements in which the iron atom(s) has given up two electrons.

fortified Having something added to it to make it better.

ion A charged atom.

mass The measure of the amount of matter in something.

matter What things are made of. Matter takes up space and has mass.

metabolism The physical and chemical properties that create energy in the body.

pole The part of a magnet where its magnetism is the strongest. There are two poles (north and south) on a magnet.

substance Any matter that takes up space.

valence The number of chemical bonds that an atom can form.

volume The amount of space that something occupies.

Further Reading

Baby Professor. *Why Does Iron Taste Funny? Chemistry Book for Kids 6th Grade*. Newark, DE: Baby Professor, 2017.

Dingle, Adrian, and Dan Green. *The Complete Periodic Table: All the Elements with Style!* New York, NY: Kingfisher, 2015.

DK. *The Elements Book: A Visual Encyclopedia of the Periodic Table*. New York, NY: DK Children, 2017.

Gray, Theodore. *The Elements: A Visual Exploration of Every Known Atom in the Universe*. New York, NY: Black Dog & Leventhal, 2012.

Websites

Elements for Kids
www.ducksters.com/science/chemistry/iron.php
Interesting facts about iron and how it is used today are featured on this website.

Fun Facts About Iron Element for Kids
www.easyscienceforkids.com/best-ironelment-video-for-kids
A six-minute video, "The Element of Iron," provides a short visual guide.

SoftSchools
www.softschools.com/facts/periodic_table/iron-facts/205
This website lists several interesting facts about the element of iron.

Bibliography

Bolm, Carsten. "A New Iron Age." *Nature Chemistry* 1: 420, 2009 (http://www.nature.com/nchem/journal/v1/n5/full/nchem.315.html?foxtrotcallback=true).

Browne, John. *Seven Elements That Changed the World: An Adventure of Ingenuity and Discovery*. New York, NY: Pegasus Books, 2015.

Coffey, Rebecca. "20 Things You Didn't Know About the Periodic Table." *Discover Magazine*, November 20, 2011 (http://discovermagzine.com/2011/nov/20-things-you-didn't-know-about-periodic-table).

Cronin, Melissa. "Neil deGrasse Tyson Explains the Origins of Atomic Elements in Our Bodies (Video)." *Huffington Post*, April 26, 2013 (http://m.huffpost.com/us/entry/3117063).

Ebbing, Darrell D. *General Chemistry*. 4th ed. Boston, MA: Houghton Mifflin Company, 1993.

Gray, Theodore. "Facts, Pictures, Stories About the Element Iron in the Periodic Table." October 3, 2017 (http://theodoregray.com/PeriodicTable/Elements/026).

Harmon, Katherine. "Get the Iron Out of Your Breakfast Cereal." *Scientific American*, May 20, 2011 (https://www.scientificamerican.com/article/get-the-iron-out-of-your-breakfast-cereal-bring-science-home).

BIBLIOGRAPHY

Heiserman, David L. *Exploring Chemical Elements and Their Compounds.* New York, NY: McGraw-Hill Trade, 1991.

Kean, Sam. *The Disappearing Spoon: And Other True Tales of Madness, Love, and the History of the World from the Periodic Table of Elements.* New York, NY: Back Bay Books, 2011.

National Institutes of Health. "Iron Dietary Supplement Fact Sheet." February 11, 2016 (http://ods.od.nih.gov/factsheets/Iron-HealthProfessional/).

Royal Society of Chemistry. "Periodic Table: Iron." Periodic Table, 2017 (http://www.rsc.org/Periodic-table/element/26/iron).

Show Me Science: Chemisty: Periodic Table of Elements. Venice, CA: TMW Media Group, 2012.

Stwertka, Albert. *A Guide to the Elements.* 2nd ed. New York, NY: Oxford University Press, 2002.

Tocci, Salvatore. *Experiments with Magnets.* New York, NY: Children's Press, 2002.

University of Nottingham. "Fe." The Periodic Table of Videos, October 3, 2017 (http://periodicvideos.com/videos/026.htm).

Weeks, Mary Elvira. *Discovery of the Elements.* Easton, PA: Journal of Chemical Education, 1968

Index

A
alloy, 4, 26, 28
atoms/subatomic particles, 9–11, 12, 14–15, 16, 20

C
compass, 7, 40, 41
compounds, 8, 31–36

E
elements, about, 6, 9, 11–15, 17–18, 20, 42

F
ferric chloride, 32, 34, 35
ferrous chloride, 32, 35–36

I
ions, 16
iron
    in food, 37–39
    humans and, 6, 30, 36, 37–39
    properties, 6, 8, 11, 13, 14, 15, 16, 17–23
    types of, 26–27
    uses for, 4–7, 8, 26, 27, 28, 34, 35–36, 39–40, 42
    where it's found, 4–6, 7, 23, 24–25, 33

M
magnetism, 6–7, 21–23, 40, 41, 42
Mars, 33
Mendeleev, Dmitry, 12
Moh's hardness scale, 19

O
ores, 24–25

P
periodic table of elements, 6, 11–16, 42

R
rust, 8, 18, 23, 27, 28, 30, 32–34

S
steel, 4, 6, 8, 20, 27, 28–30